I0815052

Metamorphic Rocks

by Grace Hansen

Abdo Kids Jumbo is an Imprint of Abdo Kids
abdobooks.com

abdobooks.com

Published by Abdo Kids, a division of ABDO, P.O. Box 398166, Minneapolis, Minnesota 55439.

Abdo Kids Jumbo™ is a trademark and logo of Abdo Kids.

Printed in China

052019

092019

Photo Credits: iStock, Shutterstock

Production Contributors: Teddy Borth, Jennie Forsberg, Grace Hansen
Design Contributors: Dorothy Toth, Pakou Moua

Library of Congress Control Number: 2018963354

Publisher's Cataloging-in-Publication Data

Names: Hansen, Grace, author.

Title: Metamorphic rocks / by Grace Hansen.

Description: Minneapolis, Minnesota : Abdo Kids, 2020 | Series: Geology rocks! set 2 | Includes online resources and index.

Identifiers: ISBN 9781532185588 (lib. bdg.) | ISBN 9781532186561 (ebook) | ISBN 9781532187056 (Read-to-me ebook)

Subjects: LCSH: Metamorphic rocks--Juvenile literature. | Rocks--Identification--Juvenile literature. | Geology--Juvenile literature.

Classification: DDC 552.4--dc23

Table of Contents

Metamorphic Rocks

Rocks are found all across the Earth. They can be buried deep underground or tower high in the sky!

Rocks can be small enough to skip on water. And they can be 30,000 feet (9.1 km) tall. Rocks are also constantly changing from one form to another.

Metamorphic rocks have been changed over time. They were once buried **sedimentary** or **igneous rocks**. They form from extreme heat and **pressure** underground.

The extreme heat comes from **magma**. **Pressure** builds from **tectonic plates** pressing together.

Tectonic plates crashing together can also cause metamorphic rock to move upward. This upward movement can create formations as large as mountains!

The Different Kinds

There are many different kinds of metamorphic rock. The kind of rock depends on how much and how long heat and **pressure** were applied. It also depends on the **minerals** that make up the rock.

Gneiss is a metamorphic rock. It has a banded appearance. It often contains lots of quartz and feldspar **minerals**.

Marble is another type of metamorphic rock. It forms when limestone is put under lots of heat and **pressure**. Marble is often used in building.

Slate forms when shale is compressed by **tectonic plates**. It is strong and good looking. So it is often used as roofing, flooring, and more.

Morphing Metamorphic Rocks

1

A sedimentary rock called shale becomes slate after it is subjected to heat and pressure

2

Higher temperatures and pressure change slate into phyllite

3

Phyllite can then turn into schist

4

After some time, schist will turn into a very hard metamorphic rock called gneiss

Glossary

igneous rock - rock formed through the cooling and solidification of magma or lava.

magma - hot, liquid matter beneath earth's surface that cools to form igneous rock.

mineral - a substance, like gold, silver, and iron, formed in the earth that is not of an animal or plant.

pressure - a steady force upon something.

sedimentary rock - rock that is formed by the accumulation and cementation of mineral or organic particles.

tectonic plate - any of the several segments of Earth's crust that move in relation to one another causing things like volcanic eruptions.

Index